I0115750

Farming that Brings Glory to God and Hope to the Hungry

A Set of Biblical Principles to Transform the Practice of Agriculture

Craig Sorley

Farming that Brings Glory to God and Hope to the Hungry:
A Set of Biblical Principles to Transform the Practice of Agriculture

Copyright © 2009 by Craig Sorley
First published 2009 by the author
First Doorlight edition 2011

Doorlight Publications
PO Box 718
South Hadley, MA 01075

All Rights Reserved

No part of this publication may be reproduced, stored in a retrieval system, or transmitted in any form or by any means – electronic, mechanical, photocopy, recording, or any other – except for brief quotations embodied in critical articles or printed reviews, without prior permission from the author or the publisher.

All Scripture quotations are taken from the Holy Bible: New International Version (NIV). Copyright © 1973,1978,1984 by International Bible Society. Used by permission of Zondervan Publishing House. All rights reserved.

Cover Design and Printing: Kul Graphics Ltd.

ISBN 978-0-9838653-0-8

DEDICATION

This book is dedicated to the farmers of East Africa.

May God bring transformation to all the farming communities in this region so that we might eliminate hunger and bring honor to the One who has entrusted this land to our care, and who created the soil, the crops, the trees, the birds and the animals.

ACKNOWLEDGEMENTS

I want to give special thanks to Dr. Harry Spaling (Kings University College), David J. Evans (Food for the Hungry International), Dr. Robert DeHaan (Dort College), Darrow Miller (Disciple Nations Alliance), and Mr. Brian Oldrieve (founder of Farming God's Way ministries), for their writings, teachings, and passion, and for their influence in shaping my thoughts on a biblical view of agriculture.

My grateful acknowledgement also goes out to the various individuals who provided help and assistance in the development of this book. Special appreciation is extended to Nelson Hard, Lee and Sue Hardman, David and Darlene Sorley, and Kent and Barb Hagen, friends at Bethlehem Baptist Church in Minneapolis, Minnesota, and to Byron Richardson, Pastor Sean Milliken, Dr. Martin Price, and Dr. Dan Fountain, who all reviewed preliminary drafts of this book and provided both encouragement and suggestions for improvement.

Special appreciation is also given to CMS Africa for their financial partnership in the initial publication of this book.

Finally I want to extend my warmest and most heartfelt thanks to my greatest advocate, my wonderful wife Tracy, for all the ways she assisted along the way. I also want to express my indebtedness to the members of our team at Care of Creation Kenya.

CONTENTS

Part 1
Introduction and Vision

Part 2
Biblical Principles to Transform
the Practice of Agriculture

PART ONE

Introduction and Vision

INTRODUCTION

Having grown up in three countries of East Africa, with the added privilege of spending many of my adult years in Kenya, I have been witness to some disturbing trends. As the gospel is more widely planted across the region, it is clear that the day-to-day hardships facing local people are intensifying. In terms of small-scale agriculture, the most common vocation in East Africa, practiced by roughly 70 percent of the population, it is the farmers engaged in this vocation who often face the greatest struggles.

Declining Harvests

Today a majority of such farmers are expressing significant discouragement over declining crop yields. Older farmers that we interview routinely report their yields are only 25-30 percent, sometimes less, of what they used to harvest when they started farming 25-30 years ago. This is understandable given the fact that many landscapes in East Africa have been continuously cropped, year after year, without rest. Compounding the problem are the rapid rates of deforestation, the erosion that follows, the intensification of farming practices due to a rapidly growing population, the growing scarcity of productive and arable land, more frequent droughts, and the increasingly erratic nature of rainfall patterns. This is an emerging and substantial problem that holds grave consequences. It is not something that will be easily or quickly resolved, and we must take stock in the implications that such trends hold for the future. Hunger is stalking the landscapes of East Africa, and this is true of other parts of the continent as well.

Tables 1 and 2 below provide a representative sample of actual figures reported to us by older farmers highlighting crop yield changes over the past 20-30 years. The Rangwe community is found in Western Kenya, near the shores of Lake Victoria, and the Tiekunu community is located a few kilometers from the edge of the Rift Valley, not too far from our office here in central Kenya. Informal surveys conducted in a variety of other locations over the past several years have shown that these are not isolated cases, but that such figures are being commonly reported by many of our agricultural communities.

Table 1: Rangwe Community

Crop	Avg. yields early 80's to early 90's	Avg. yields 2005 to present	% yield compared to early 80's & 90's
Maize	12 bags/acre	4 bags/acre	33%
Sorghum	6 bags/acre	1 bag/acre	17%
Beans	1 tin sown yielded 20 tins	1 tin sown yields 6 tins	30%

Table 2: Tiekunu community

Crop	Avg. yields early 80's to early 90's	Avg. yields 2005 to present	% yield compared to early 80's & 90's
Maize	30 bags/acre	7 bags/acre	23%
Beans	20 bags/acre	5 bags/acre	25%
Potatoes	100 bags/acre	10 bags/acre	10%

*Note: In rural Kenya large gunny sacks are commonly used to measure yield. One bag of maize weighs 90 kilograms. 30 bags/acre for example, translates to 2.7 metric tons/acre.

Discouraged Farmers

Apart from the actual figures on crop yields, there is a corresponding cultural and social trend that must also be recognized. Strong sentiments of discouragement are being passed on to the next generation. As the struggles of small-scale farmers intensify, young people are becoming increasingly disillusioned with the prospect of becoming farmers themselves. Here in Kenya, it is virtually a cultural norm for parents to urge their children to get a "good education," followed by the accompanying admonition that says, "I hope you don't have to become a poor farmer like me!" Our cultural worldview in East Africa, and across the rest of the continent, often condemns farmers, placing them at the bottom of the social ladder. In the minds of many, farming is increasingly seen as a fruitless and futile way of life.

Dignity Restored

This kind of worldview is not only dangerous, it is decidedly not biblical. Neither is it hopeful as Christian leaders, local farmers, missionaries, or other segments of society consider the challenging task of feeding future generations. There is a great deal at stake here. How farmers care for their land, or how they do not care for their land, holds tremendous repercussions for future food supplies. Even those who live and operate in urban communities must still eat, on a daily basis, and the very foundation of the economic activity that takes place within our cities or towns is built upon and supported by a vital and necessary connection to agriculture. If farmers have such a low view of themselves, reinforced by their society, and if they cannot take pride in what is still the most common vocation on the continent, then hope grows faint. Those within the body of Christ can no longer afford to neglect this issue. We must restore to farmers the dignity and respect that they deserve, as people created in God's image. If we cannot substantially encourage and equip farming cultures to care for their farming landscapes in the best way possible, then we must accept the inevitable outcome. The specter of hunger will continue to stalk this continent with growing intensity.

As the combined circumstances facing small-scale farming communities across Africa become more difficult, some critical questions must be raised for those of us who claim Christ as Lord.

- How does God fit into this picture of agriculture?

- Does the Bible have anything to say about farming?

- Why are agricultural systems, and the poor in Africa, facing such hardships when Christianity is growing by leaps and bounds in many of these regions?

- If the gospel can heal and transform people, can it also heal and transform the very farming systems that feed us?

For several years now, as we have worked together in Kenya, my counterpart Francis Githaigah and I have made it common practice to pose the following question to our Christian brothers and sisters, and particularly to farmers:

"What does your faith in Jesus Christ mean
for your way of life as a farmer?"

If I am correct, this question puzzles you as the reader in the same way that it puzzles the average farmer who is a member of a local church. In almost every case there are two typical answers that we receive to this inquiry. The first is usually a look of bewilderment, followed by a statement along these lines: "I have never considered such a question before." The second response is often an honest confession that simply admits "I don't know what my Christian faith means for farming."

Such a statement should come as a shocking one to us who call ourselves Christians, and this is precisely where we as believers must recognize a great tragedy. While agriculture is the economic mainstay for the majority of communities across Africa, most Christian farmers have little or no meaningful connection between their faith in Christ and their primary vocation in life.

Part of the responsibility for this lies with those of us who have come as missionaries. We have brought the gospel to a continent of farmers, and that gospel has typically said little or nothing about farming. But part of the responsibility also lies within those of us who have accepted that gospel, and in our own failure to integrate the gospel into the very fabric of our lives. What is taught and learned on Sunday mornings in church often has little or no bearing on what is practiced in the field during the rest of the week. If the gospel has not been shown to be relevant for what is the most common way of life on the continent, we must recognize that something very significant has been missing from this picture.

The purpose of this book is to put God back into the center of agriculture. He is the First Gardener, the First Farmer, and the First Forester, and we have left Him almost entirely out of this important dimension of life. It is time that we put Him back into His rightful place. God's word can teach us a great deal about farming. And His call to Christians that we minister to the poor should be reflected in farming systems that give priority to minimizing the suffering of the poor.

If Christianity is taking root in a place like Kenya, then surely something should be taking root in our farming landscapes as well. Our gardens and farms should bear tangible witness to the fact that we are committed to Christ, just as the rest of creation bears tangible witness to the beauty of our Lord. If the gospel can heal and transform people, then surely the gospel can heal and transform how we farm. This kind of renewal in thought and practice honors the truth we see in Romans 12:2.

> Do not conform any longer to the pattern of this world, but be transformed by the renewing of your mind. Then you will be able to test and approve what God's will is - his good, pleasing and perfect will.

This book is about not conforming to the patterns of this world; it is about being changed by the Gospel of Jesus Christ and through the Scriptures in our attitudes and thinking about farming. A major shift in our perspective is needed. We need to move from a mindset focused merely on maximum yields and maximum profit to a mindset focused on stewardship, compassion, and care. A God-centered and biblical ethic must be our first and foremost guide in all matters pertaining to agriculture. The end goal is to both convict and inspire all Christians to make the changes that are necessary, so that we might glorify God in the practice of agriculture and bring hope to the hungry.

A Vision for
Agriculture and Farmers

For those who call ourselves Christians, whether we live in a big city in the developed world, or in a small village in Africa, our commitment to Christ calls us to vigorously embrace a concern for hunger and poverty, wherever it may occur. In today's world we must ask ourselves some tough questions. Are we content, for example, with the reality that millions of people in countries like Ethiopia, Somalia, Kenya, and Tanzania face serious food shortages on a semi-regular and often chronic basis? As farmers or citizens of these countries, as foreign missionaries who minister and work there, or as believers who live across the oceans, are we willing to simply accept this growing reality as we look into the future? Are we content with simply providing relief supplies to the hungry when drought or famine strikes?

As challenging as these questions are, they point us to an even more fundamental question that must be asked.

"Do we have a God-centered vision for agricultural land?"

When we look at a community, with the hope to see lives submitted to Christ and churches planted, does the vision end there, or does it embrace something bigger? Does that vision incorporate the land that sustains the people of that community? Have we taken the next step to develop a plan for equipping those people with the tools necessary to restore their land, so that they might enjoy the dignity of feeding their own families with their own farm-grown resources? Have we made a deliberate effort to lay this issue at God's feet, requesting that He provides us with a comprehensive biblical vision for farming and for eliminating hunger?

If we examine the priorities of the church today, and if we are honest with ourselves, it is only reasonable that we confess the following reality: The vast majority of us within the body of Christ have little or no vision on this topic. And there is a consequence to this lack of vision. Unfortunately it translates to less hope for the hungry. It is time, therefore, that we develop and implement such a vision.

The reason for this book, and the goal of Care of Creation Kenya, is to encourage all Christians to join together in promoting a sound biblical vision for agricultural stewardship. The passage below from Exodus is a true reflection of the dream that we hold for the farmers of Africa. In keeping with this example set by our Heavenly Father, we want to acknowledge the suffering and hardship that currently affects many of our communities. We want to demonstrate a Christ-like concern for them. And we want our work to help restore the damage that has been done, so that people can once again enjoy the blessings of a healthy and productive landscape.

The Lord said, 'I have seen the misery of my people in Egypt. I have heard them crying out because of their slave drivers, and I am concerned about their suffering. So I have come down to rescue them from the hands of the Egyptians and to bring them up out of that land into a good and spacious land, a land flowing with milk and honey...' (Exodus 3:7-8)

The fruit of a sound biblical vision for agricultural stewardship

As we consider the task at hand, and the challenges that lie ahead, we recognize that transforming and healing the landscapes of Africa must begin with transforming the hearts and minds of people. Only through God, and through the power of His word, can a true transformation of the heart take place. As Jesus reminds us in Matthew 7, "*You shall know them by their fruit.*" Lives changed by Christ should reflect real changes in attitudes and behaviors. How does such change manifest itself in the life of a farmer?

As one example, the prayer written below could serve as an illustration. It demonstrates a farmer whose mind has been transformed by Christ, who is thinking biblically about agriculture, and who is seeking to bear fruit in both the spiritual and physical dimensions of his life. This represents our hope for how God can change the farmers and farming communities of Africa.

Lord I want to honor you today as I work this land that you have entrusted to my care. Give me wisdom as I strive to make this part of your creation even more bountiful. Enable me to enrich this land for my family, for your creatures, and for future generations. Teach me how to share the love of Christ in both word and deed.

The prayer of a Godly Farmer

PART TWO

Biblical Principles to Transform the Practice of Agriculture

Principle 1

A vision for transforming agriculture begins with God, and His intention that the earth would be a place of abundant provision and blessing to His people

As we consider the urgent need for developing a biblical vision for agricultural stewardship, the most appropriate starting place is to examine what God created in the beginning. When we look at the creation story in Genesis 1 we see a beautiful picture of the earth being filled with an abundance of life. The waters teemed with fish, the skies with birds, and the land was filled with plants and animals. God intended His creation to burst forth with life as a demonstration of His majesty and glory. After completing His work God reflected on what He had created, and He was delighted with what He saw. The creation, as God made it, was a place of abundance that brought Him pleasure.

> Then God said, 'Let the land produce vegetation: seed-bearing plants and trees on the land that bear fruit with seed in it, according to their various kinds.' And it

was so. The land produced vegetation: plants bearing seed according to their kinds and trees bearing fruit according to their kinds. And God saw that it was good. (Genesis 1:11-12)

And God said, 'Let the water teem with living creatures, and let birds fly above the earth across the expanse of the sky.' So God created the great creatures of the sea and every living and moving thing with which the water teems, according to their kinds, and every winged bird according to its kind. And God saw that it was good. God blessed them and said, 'Be fruitful and increase in number and fill the water in the seas, and let the birds increase on the earth.' (Genesis 1:20-22)

God saw all that he had made, and it was very good. (Genesis 1:31)

When we move on to Genesis 2 we see a very similar picture, and this time it reveals an important clarification on God's original intentions for mankind. Adam and Eve are to enjoy a life intimately connected to the land. They are placed into a marvelous garden, filled with many species of trees, and Scripture clearly points out the responsibility they are given to care for this paradise. They are privileged to enjoy the fruits of creation, and they also enjoy a rich and abundant relationship with the Creator Himself. The Garden of Eden, as God made it, was a place of abundant provision, both physically and spiritually.

Now the Lord God had planted a garden in the east, in Eden; and there he put the man he had formed. And the Lord God made all kinds of trees grow out of the ground - trees that were pleasing to the eye and good for food. In the middle of the garden were the tree of life and the tree of the knowledge of good and evil. A river watering the garden flowed from Eden... (Genesis 2:8-10)

The Lord God took the man and put him in the Garden of Eden to work it and take care of it. (Genesis 2:15)

This story of beauty and abundance does not end in Eden, even though a significant change takes place when Adam and Eve rebel against God, and they are banished from the garden. We find that this theme features very prominently later on in the Old Testament when we read the story of God's covenant with Abraham. God approaches Abraham and instructs him saying, "Leave your country, your people... and go to the land I will show to you...I will make you into a great nation...and all peoples on earth will be blessed through you." (Genesis 12: 1-3).

God's covenant with Abraham would lead to the greatest blessing ever bestowed on mankind, the coming of the Christ child, born through Abraham's lineage. But that was only part of the blessing and part of the covenant. There was another important component. Prior to Christ's coming the nation born from Abraham's family was to receive a significant gift in the form of the Promised Land. In Exodus 3 we see a stunning description of what that land is like. It is a "good and spacious land, a land flowing with milk and honey" (vs. 8). And later on in Leviticus 14:7, after Joshua and Caleb returned from their excursion to investigate the territory, they reported the following: "The land we passed through and explored is exceedingly good."

Now if we pause to contemplate this for a minute, we could all agree on something. If at some point in time we were able to witness a river flowing through our community, filled with either milk or honey, such a day would go down as an unforgettable event in history. Now Joshua and Caleb did not see actual rivers of milk and honey, but they certainly affirmed the image God had painted with His words. This image is still known across the world today, even by the biblically illiterate. It is an unforgettable image! God's intention is to bless the Israelites, and the picture of abundance that emerges here could not be more striking.

But there is an important condition. Such abundance is not granted freely and without responsibility. It is dependant upon obedience to the One who created and provides the blessing. In Deuteronomy 8 and 11, God's instructions are very clear.

> Observe the commands of the Lord your God, walking in his ways and revering him. For the Lord your God is bringing you into a good land - a land with streams and pools of water, with springs flowing in the valleys and hills; a land with wheat and barley, vines and fig trees, pomegranates, olive oil and honey; a land where bread will not be scarce and you will lack nothing... When you have eaten and are satisfied, praise the Lord your God for the good land he has given you. Be careful that you do not forget the Lord your God, failing to observe his commands... (Deuteronomy 8:6-11)
>
> So if you faithfully obey the commands I am giving you today - to love the Lord your God and to serve him with all your heart and with all your soul - then I will

send rain on your land in its season...so that you may
gather in your grain, new wine, and oil. I will provide
grass in the fields for your cattle, and you will eat and
be satisfied. (Deuteronomy 11:13-15)

Based on these passages what lessons can we take from
them for the purposes of agriculture? What we have seen
is a vivid picture of abundance in creation, in the Garden
of Eden, and in the Promised Land, as long as the Israelites
remained faithful to God. So as a first and most important
lesson, we as Christians should fix our eyes upon these
images, as examples given by our Lord, as goalposts to guide
us as we pursue a biblical vision for agricultural stewardship.
In a world where the abundance of the land is fading, God's
people should be working to restore that abundance.

When we look to the hills, and see that the forests have
been cut down and not replanted, we must have a vision for
replanting them. When we look at our rivers, and see them
filled with sediment, and tainted by chemical pollutants
that have washed from our fields, making the fish and water
unsuitable for human consumption, we must have a vision
for seeing those rivers flow clean once again. When we see
that our crop yields are in decline, and when we examine
our topsoil and realize that much of it has washed away, we
must have a vision for restoring the fertility of that soil.

Secondly, and just as important, as our vision develops it
needs to be accompanied by a comprehensive worldview
based on the scriptures; a worldview which will help us to
identify specific ways in which we can demonstrate a true

commitment to honor God in our practice of agriculture. Such a worldview, and how we express it, should include our thoughts, our attitudes, and our actions on the ground. This is how we can glorify God, and bring hope to the hungry, in the practice of agriculture.

And thirdly, as we enter into this process, we must turn to God with a repentant spirit and cultivate a consistent habit of coming before Him in prayer. We must willingly admit that the infertility and erosion on our landscapes are problems that we have brought upon ourselves, due to our own neglect and carelessness. We cannot blame God for the difficulties we face for He is the One who originally created such a beautiful world of abundance. Our vision must begin with God, and as broken vessels still prone to wander and make mistakes we must continually depend upon God as we strive to carry out that vision. A prayer that communicates a contrite spirit, a commitment to do what is right, and a dependence upon the One who can make it all possible might be represented as follows:

> Lord, I confess I have been negligent in caring properly for the land you have entrusted to me, and I recognize this has undermined my witness for Christ. Please equip and enable me to glorify you as I strive to restore the abundant provision that once grew from this land.

A vision for transforming agriculture begins with God, and His intention that the earth would be a place of abundant provision and blessing to His people

Principle 2

How we do agriculture should glorify God and reflect our commitment to Christ

A number of years ago, Rob De Haan, a professor of agriculture at Dordt College, Iowa, posed an interesting question. If we were to embark on a tour of agricultural landscapes in the U.S., observing only basic farm operations and the landscapes themselves, and without actually speaking to the farmers themselves, would we notice discernable differences between farms that were run by Christians and those run by non-Christians? Could we identify which farms were operated by followers of Christ, based on such observations? Would there be evidence, on the ground or in the dairy barn, to help us determine which systems were being managed by farmers committed to our Lord, and which systems were being managed by unbelievers who viewed life from a strictly secular perspective?

The conclusion of the group at Dordt College was that such a task would be virtually impossible. Not in every case, but for the most part, Christian farmers do nothing substantially different than their non-Christian counterparts. Generally speaking, farming is farming. It has nothing to do with your spiritual commitments, and that's the end of the story.

We have been asking the same question of our Christian farmers here in Kenya for several years now, and the response is very much the same. Christians willingly admit there is no discernable or substantial difference between how they farm and how their non-Christian counterparts farm. But when pressed to think more carefully, they readily admit that there should be a difference. The truth that there should be a discernable difference can be quickly drawn out by presenting the two diagrams below and by posing the two simple questions that accompany them.

by Terry Beck

by Terry Beck

Which community is the Christian community?

Which community is honoring God in their use of the land?

In every case where we have presented these images to Christians they quickly and almost instinctively select the second image. While most of them have never considered such a question before, their swift response reveals a glimmer of hope. In their heart of hearts they know that God expects something better. Yes, it is true, God can be glorified in how we farm. The manner in which we do our farming can serve as a very real reflection of our commitment to Christ.

The biblical foundation for this principal comes from I Corinthians 10:31.

> So whether you eat or drink or whatever you do, do it
> all for the glory of God.

Notice the completeness and all encompassing nature of this passage. Whatever we do in life, whether it is something as mundane and simple as eating a meal, or as significant as bringing relief supplies to the starving, it should all reflect the goodness and greatness of God. In the agricultural context we could perhaps reword this verse to say the following: *"So whether you grow vegetables or grain or produce orange juice, do it all for the glory of God."*

As we consider building a biblical vision for agricultural stewardship, this is one of the first and most important questions we must ask. Do our agricultural systems reflect the goodness and greatness of God? Do they speak of His excellence and perfections in how they are managed and in the proper care that is given to the land? If they do not, then we obviously need to make some changes.

The differences that we see in the second image - the contouring on the slopes to prevent erosion, the forest still intact on the hillside, the trees planted in the village and in the fields, the pasture that is still in good condition, and the pond and the stream that still flows - all help to illustrate a simple but very fundamental concept that we must embrace. The beauty of a healthy, productive, and well cared for agricultural landscape can be a testimony to the beauty of people whose lives have been changed by Christ.

When other farmers and those who are hungry see such examples of excellence, they will want to know what is going on. They will be attracted and drawn to these communities by the outer beauty manifested on the land, only to discover there is also an inner beauty to be found in the people who live there. Indeed many opportunities will arise to either disciple such farmers or to introduce them for the very first time to Christ. This highlights the ultimate goal that we should be striving towards. It brings into focus what should be central to our thinking and practice as Christians operating in the realm of agriculture, for Christ is the One who can transform both people and the landscapes that sustain them. If the gospel can heal and change people, then surely it can heal and change how we do farming.

> How we do agriculture should glorify God
> and reflect our commitment to Christ

Principle 3

Farming is a meaningful and noble way of life because God was the First Farmer

In the introduction to this book we underscored the discouragement that confronts many of our farming communities today. What normally accompanies and reinforces that discouragement is a worldview which often looks down upon farming as a way of life. Instead of being recognized for the important role that they play, farmers are often placed at the bottom of the social ladder. While this may not be true in every corner of the earth, it is certainly true here in Africa and in many other parts of the underdeveloped world.

How can dignity and a sense of self-respect be restored to the millions of small-scale farmers who live and operate under this kind of unhealthy atmosphere? Undoubtedly this represents one of the greatest challenges and one of the most important wholesale changes that must take place. If we are going to completely transform agriculture, if we are going to successfully bring healing to our farming systems, and if we are going to adequately feed future generations, we must pay close attention to how we view farming as a way of life. Nothing could be more critical than to examine how farmers perceive themselves, and how the rest of society perceives farming.

We must ask some very tough questions. Is farming an important vocation? Is it a respectable way of life? Is it recognized as a critical endeavor, essential to the future well-being of the nation? Most importantly, does the church communicate a God-centered and biblically-based message that affirms and encourages farmers?

The unfortunate but honest reality is that we generally show little respect to our farmers and little appreciation for the hard work that they do. When farming is increasingly viewed as a demeaning or futile way of life, when it is characterized by discouragement and a loss of hope, we need to recognize the long-term and potentially devastating consequences of such trends. Casting this type of view upon agriculture is certainly not biblical, and this is where God must enter the picture. His word can bring much needed transformation.

Let us see how that transformation can begin by examining two key passages in Genesis.

> Now the Lord God had planted a garden in the east in Eden; and there he put the man he had formed. (Genesis 2:8)

> The Lord God took the man and put him in the Garden of Eden to work it and take care of it. (Genesis 2:15)

27

Genesis 2:8 holds a golden nugget of truth that can usher in one of the most necessary shifts in worldview, the very key that can open the door to transforming the minds of hopeless farmers. God was not someone who just walked across the red carpet after someone else had designed and planted the garden. He himself was the One who planted the garden. And what a garden it must have been! A true masterpiece that was glorious beyond our imagination. There is another word that we often use interchangeably for someone who plants a garden. We identify such a person as being a farmer.

There is something important here that most of us have never considered before. In the context of agriculture, when we reflect on this story in Genesis, God emerges before us as being the First Farmer. He planted the very first garden, and thus, He is the very First Farmer. This brings us face to face with some profound questions. What is God's view of farming? Does He view farming as a demeaning and useless way of life? Or does He take the soil of the earth into His hands and look upon farming as a noble vocation to be pursued with excellence? The conclusion is not only clear, it is liberating and transforming. Since God is the Model Farmer, the One who sets the standard, we can now begin unfolding a brand new picture on this topic that holds far-reaching implications.

The First Farmer principle is profoundly powerful because it connects us with God, and it fills an enormous void. Instead of conforming to the social pattern that looks at this picture as an empty cup, the Bible helps us to see the vocation of farming in a new light, as something that overflows with a significant sense of legitimate meaning and dignity. God our Provider planted a garden for Adam and Eve's benefit, and in similar fashion, human farmers coax food from the earth to feed themselves and others. Indeed this is a worthy profession. God's garden was a work of art that emanated His excellence, and it serves as the premier example to be followed by farmers who are made in His image.

But that is not all. When we remind ourselves once again of what is stated in Genesis 2:15, even more significance is added to this transforming concept. After God worked to create His matchless garden, He then commissioned Adam to take care of it.

Now imagine for a minute if the president of your country showed up at your doorstep one day with a request. "I have a job for you" he said. "I have decided that you should take charge over the management of my presidential garden." Give some serious consideration to such a scenario. How would you respond? For most of us our hearts would start beating faster, we would get nervous, and we would wonder to ourselves, "Am I really prepared and ready to take on such an important task for the president?"

So from a biblical perspective we see that God gave Adam a very important job. His commissioning of Adam to care for the Garden of Eden multiplies the importance of the job far beyond any task that could be assigned to us by even the most important of earthly leaders.

So God was the First Farmer, and tending the garden was mankind's first job description. This was God's original intention for mankind from the beginning, and this is GOOD NEWS for the farmers of Africa. They can take courage in knowing that their job is a noble one, that their task is one of significance, and that God has given them a special responsibility. For those of us who are not farmers, and especially those who are pastors or church leaders, we need to build this type of worldview into our farming communities. We need to communicate a respect for farmers that reinforces a biblical perspective. Both now and into the future, people from all walks of life will benefit when the farmers of today glorify God by taking care of their landscapes in the best way possible.

> Farming is a meaningful and noble way of life
> because God was the First Farmer

Principle 4

Godly agriculture gives priority to healthy food and to the needs of the poor and hungry

I was travelling in Tanzania a couple of years ago and happened across an interesting billboard advertising cigarettes. The image showed an attractive young lady with a *Sweet Menthol* cigarette pack opened on a restaurant table. She was arm wrestling a handsome young man at this table, and was overpowering him. The slogan on the side of the billboard read "Ni freshi, ni poa, ni yako!", which translated to English means "It's fresh, it's good and it's yours!" The obvious message was that it was a fresh and good thing to smoke *Sweet Menthol* cigarettes, and that doing so would help women triumph over their battles with men.

Now apart from recognizing the insidious falsehood and ungodly nature of this advertisement, it began to take on additional meaning as I continued my journey. There I was, in central Tanzania, where local farming communities were facing years of chronically low rainfall, which led to chronic hunger. So on the one hand, people were suffering from consistently poor crop yields, and on the other hand, they were being persuaded to spend what little money they had on tobacco, a commodity crop that uses up valuable agricultural land to produce something harmful to human health. Those involved in marketing hazardous cigarettes, and those who succumbed to the temptation to purchase them were equally guilty in perpetuating a situation that contributed to the detriment of the land and its people.

The question I want to raise here is this: Where do our priorities lie when it comes to the use of agricultural land? Is our first priority to play global economics, and to grow as many cash crops as possible to make as much profit as possible, regardless of how it affects the land or its people? Or do our agricultural systems glorify God by giving first priority to meeting the long term nutritional needs of the poor and hungry?

We do not need to perform an in-depth analysis to realize that the first scenario is far more frequent than the second. For the most part, agricultural systems across the globe cater to the demands of the affluent and the wealthy. Achieving real food security, the consistent provision of adequate food supplies for all people, is not the first priority in many of our modern practices, which have turned the livelihood of farming into industrial agribusiness. To get our priorities straightened out, we need to go back to scripture and learn from God.

In examining the passages below, let us take note that God is speaking to an agriculturally based nation. His expectation of the Israelites is that they express God's benevolence in their practice of agriculture.

> When you reap the harvest of your land, do not reap to the very edges of your field or gather the gleanings of your harvest. Do not go over your vineyard a second time or pick up the grapes that have fallen. Leave them for the poor and the alien. I am the Lord your God. (Leviticus 19:9-10)

> When you are harvesting in your field and you overlook a sheaf, do not go back to get it. Leave it for the alien, the fatherless and the widow, so that the Lord your God may bless you in all the work of your hands.... Remember that you were slaves in Egypt. That is why I command you to do this. (Deuteronomy 24:19-22)

> When you have finished setting aside a tenth of all your produce in the third year, the year of the tithe, you shall give it to the Levite, the alien, the fatherless and the widow, so that they may eat in your towns and be satisfied. (Deuteronomy 26:12)

Just as there was a striking beauty that emerged from God's model garden in Genesis 2, there is a striking beauty that emerges from these passages in Leviticus and Deuteronomy. God intended the Israelites to use their agricultural systems in a manner that embraced the needs of the poor. While the Israelites were able to rejoice in bringing in their harvest, God called them to be generous, and to be constantly mindful of the unfortunate. The intensity of their suffering in Egypt was to be remembered, so they would never make the mistake of forgetting the poor by hoarding their blessings to themselves.

The author of Proverbs in chapter 21 verse 13 gives a stark warning to those who forget the poor.

> If a man shuts his ear to the cry of the poor, he too will cry out and not be answered.

Do our agricultural systems today often shut their ears to the cry of the poor? Unfortunately the answer is yes. Either little effort is given to empowering the poor in their own agricultural needs, or farm laborers are given marginal compensation for the work performed in our large scale industrialized operations. In the Scriptures God is pointing us to something radically different.

A nation dedicated to Godly agriculture gives first priority to the feeding of its people and to producing adequate supplies of healthy food. Commercially grown specialty crops for the wealthy should take second place. Due to the growing scarcity of arable land, neither should we devote our land to the cultivation of any crop that is addictive or harmful to human health, or harmful to the land itself. And as we consider what would bring the greatest glory to God, we should not be satisfied with merely providing handouts to the hungry. We should be committed to a process of transforming our agricultural systems in a way that lifts people up, and that equips and enables the poor to one day feed themselves.

> Godly agriculture gives priority to healthy food
> and to the needs of the poor and hungry

Principle 5

Godly agriculture cares for the whole creation and benefits the whole community of life

While the admonition in Scripture clearly encourages us to pay close attention to the poor, we must also recognize that such admonition is part of a bigger picture. The practice of agriculture takes place not in isolation, but within the larger context of creation. As we will see in Scripture, God had broad intentions for the Israelites in how they were to manage the Promised Land. Their stewardship was meant to extend benefits beyond that which would be enjoyed by people alone.

If we were to examine most agricultural systems in Africa today, what would we conclude about the impact that such systems are having on other parts of creation? Do our current systems generally result in further degradation of other God-given resources, or do they operate in a healthy balance that sustains the integrity of those resources? Unfortunately, when we examine the evidence, the first scenario is far more accurate of reality as compared to the latter. Many land-use practices employed today are leading to the degradation of other resources. Some well documented examples include:

- Deforestation and the conversion to agriculture of designated forest lands, which were originally set aside to protect catchment areas, rivers, streams, and other water resources.

- The cultivation of sloping land with little or no soil conservation measures, facilitating erosion, flash floods, lower crop productivity, and the further deterioration of streams, rivers, and springs.

- The elimination of most indigenous plant and tree communities, disrupting the natural balance of indigenous birds and insects that often help to control crop pests.

- The careless use of pesticides leading to numerous health risks for humans, livestock, the contamination of fisheries, the loss of beneficial honey bees, etc.

- The long-term over-use of both fertilizers and pesticides leading to the contamination of drinking water supplies in wells and aquifers.

- The intentional use of agricultural pesticides for exterminating wildlife.

If such practices are common to life in many of our agricultural communities today, how should a God-centered approach to agriculture respond to such realities? Let us first examine what we see in Scripture.

> Speak to the Israelites and say to them: 'When you enter the land I am going to give you, the land itself must observe a sabbath to the Lord. For six years sow your fields, and for six years prune your vineyards and gather their crops. But in the seventh year the land is to have a sabbath of rest, a sabbath to the Lord. Do not sow your fields or prune your vineyards. Do not reap what grows of itself or harvest the grapes of your untended vines. The land is to have a year of rest. Whatever the land yields during the sabbath year will be food for you—for yourself, your manservant and maidservant, and the hired worker and temporary resident who live among you, as well as for your livestock and the wild animals in your land. Whatever the land produces may be eaten.' (Leviticus 25:2-7)

While I intend to deal with the concept of the sabbath rest in another section, I want to set that aside for the moment and highlight the total picture portrayed in this passage. Notice the all encompassing nature of God's unmistakable concern for this agricultural community. We have field crops, vineyards, Israelites, servants, temporary workers, foreigners, domestic livestock, and wild animals represented. God is very concerned that the land is not abused or exhausted, but that proper stewardship would bring benefits to all dimensions of His creation, both human and non-human.

Now we don't have forests or rivers or other aspects of creation specifically listed in this passage, but the fact that livestock and wild animals are mentioned gives us a strong clue.

Such animals were clearly meant to exist, including the water, food, and habitat required to sustain them. Agriculture, as God intended, was obviously meant to fit into the larger context of God's creation, benefiting the whole community of life.

This raises some tough questions regarding the problems listed earlier. If our agricultural practices are causing soil erosion and reducing the productive potential of the land, if they are leading to a degradation of water resources, if they are eliminating the plant, animal, and bird communities that God originally put in place, and if they are posing risks to human health, is God being glorified? Is our agricultural system caring for the whole creation as portrayed in Leviticus?

It is appropriate here to introduce another passage of Scripture as we consider how God would want us to handle such challenging questions.

> As for you, my flock, this is what the Sovereign Lord says: I will judge between one sheep and another, and between rams and goats. Is it not good enough for you to feed on the good pasture? Must you also trample the rest of the pasture with your feet? Is it not enough for you to drink clear water? Must you also muddy the rest with your feet? Must my flock feed on what you have trampled and drink what you have muddied with your feet? (Ezekiel 34:18-19)

In the larger context of this passage, we learn that God is expressing his outrage that the "shepherds" of Israel have failed in their task to properly care for His "flock" (see vs. 2).

These "shepherds" were the spiritual leaders who had been appointed to guide and help the people, and they had used their power and influence to selfishly bring gain to themselves, at the expense of those underneath them. They had squandered and abused the blessings they were privileged to enjoy and now God is judging them for such abuse. In doing so He uses a powerful analogy from agriculture to make His point.

We can take at least two important lessons from this passage. The first is that we as Christian leaders will ultimately be held accountable by God for the position that He has granted to us in guiding and shepherding His people. In the context of this book, we cannot avoid the following question: Have we been guiding and shepherding our people in the area of a biblical approach to environmental and agricultural stewardship?

The second lesson is more down-to-earth. God used the analogy of trampled pastures and muddied waters because the Israelites, an agrarian culture, could easily understand that such mistreatment of resources was a gross violation. A powerful rebuke is given to anyone who sins against his fellow man by trampling resources and leaving behind a land that is ruined.

While it is true that we don't have quick and easy answers to all of the problems listed earlier, knowing that sometimes we may need to use fertilizers or pesticides, and that sometimes we may need to clear land for agriculture use, etc., we should only engage in such activities with a resolute determination to exercise utmost care. We should carry out such efforts with a solid commitment to minimize potentially negative impacts, recognizing that certain boundaries should not be crossed. Any activity that leads to a degradation of the land should be brought into question and scrutinized. Such striving to maintain the quality and productivity of the land fulfills the quest to glorify God in our practice of agriculture. There is also an important corollary to this concept. If Godly agriculture cares for the whole creation and benefits the whole community of life, Godly agriculture will also strive to restore land that has been abused, damaged, or rendered unproductive.

> Godly agriculture cares for the whole creation and benefits the whole community of life

Principle 6

Like God's garden, our farming systems should be diverse, incorporating many kinds of trees, plants, crops and animals

When I was a young teenager our family moved to the small town of Mbale, Uganda, where my father undertook endeavors to minister to people through community-based health care. Shortly after our arrival we learned that a devastating drought had recently ended. During the drought many people had nearly perished from hunger, but something about their agricultural system had prevented tragedy. Mango trees dotted the farms, and as the crops withered and died, these trees withstood the drought, and they produced fruit. The people survived that year because they had mangoes to eat.

Agricultural experts around the world have always encouraged farmers to not depend solely upon just one or two crops. The best defense against unexpected pest invasions, drought, or other tragedies is to develop a farming system that is diverse. If an insect comes along and completely destroys your corn crop, then at least there are other crops, trees, or animals which can bring a family through the lean times. Where do we first see the concept of diversity being heralded as a good and beautiful thing? Let us reflect again on the very beginning of Scripture.

> Then God said, 'Let the land produce vegetation: seed-bearing plants and trees on the land that bear fruit with seed in it, according to their various kinds.' And it was so. The land produced vegetation: plants bearing seed according to their kinds and trees bearing fruit with seed in it according to their kinds. And God saw that it was good. (Genesis 1:11-12)

> Now the Lord God had planted a garden in the east in Eden, and there he put the man he had formed. And the Lord God made all kinds of trees grow out of the ground - trees that were pleasing to the eye and good for food. In the middle of the garden were the tree of life and the tree of the knowledge of good and evil. (Genesis 2:8-9)

> You are to bring into the ark two of all living creatures, male and female, to keep them alive with you. Two of every kind of bird, of every kind of animal and of every kind of creature that moves along the ground will come to you to be kept alive. You are to take every kind of food that is to be eaten and store it away as food for you and for them....Take with you seven of every kind of clean animal, a male and its mate, and two of every kind of unclean animal, a male and its mate, and also seven of every kind of bird, male and female, to keep their various kinds alive throughout the earth. (Genesis 6:19-7:3)

Do we serve a God who delights in biological diversity? Do we serve a God who is concerned about protecting and preserving such diversity? These Scriptures answer such questions with a resounding "Yes"! God is very pleased, in Genesis 1, to witness the variety of tree and plant species that he had created. In chapter 2, God fashions the Garden of Eden to be filled with many kinds of trees that served to provide both beauty and nourishment. Later on, in the story of Noah, God makes it abundantly clear that every kind of animal and bird was to be kept alive. In fact, the phrase "*every kind*" and similar terminology is used more than fifteen times between chapters 7 and 9 of Genesis.

Around the world today scientists, environmentalists, and agronomists correctly bemoan the loss of biodiversity, and unfortunately, they often receive only token affirmation from Christians in their conservation efforts, or in their attempts to rescue a species from extinction. Yet right under our noses the Scriptures plainly demonstrate God's own affinity for diversity.

One positive concept that is gaining more attention around the world today is the idea of agroforestry, which brings the disciplines of agriculture and forestry together with conservation practices. It is a deliberate attempt to maximize a variety of food production and environmental benefits that can be created when trees are combined with crops and/or animals. This usually results in a system that is diverse, more productive, and sustainable.

Agroforestry can also be of tremendous benefit for women, especially in Africa, where growing numbers are burdened with rather desperate situations, travelling further and further each year to collect ever dwindling firewood resources. In a well-maintained agroforestry system, this burden can be eased or eliminated altogether, allowing a woman to gather most if not all of her firewood right from the farm itself. This permits her to spend valuable hours in the more important activities of caring for the land and her family.

From a biblical perspective, I like to point out that agroforestry is not a new concept created by man. It is a system that was modeled for us in the Garden of Eden. Trees were a major feature in God's garden, and in fact, the tree was ordained by God to serve as a symbol of life. We see that the tree of life itself stood in the very middle of that garden. As we consider the types of species that might be included in our farms we must also recognize that God, in His wisdom, has created many different "*kinds*" of trees, all of which are designed to flourish and thrive in different places and environments. A farming system that seeks to incorporate native species of plants and trees as much as possible truly honors the One who originally placed many "*kinds*" of plants and trees to live and exist in that area.

When we travel across East Africa today, plenty of evidence exists that our agricultural systems have lost much of their diversity. Corn is now the predominant crop, and drought tolerant foods like millet, sorghum, and cassava are no longer being grown in most areas. A couple of years ago we were working with a group of Baptist pastors from the Nyeri region of Kenya, who also happened to be farmers. One of their main complaints was the growing trend of insect infestations, and the fact that they had to use larger amounts of costly and dangerous pesticides to control such infestations. We asked them why they felt this problem was worsening, and upon reflection, they gave a very perceptive answer.

> The indigenous trees and forests that used to be common in this area have all been cut down. There used to be birds and insects living in those forests that would come and eat the pests on our crops. Now those birds and beneficial insects are gone.

God's intricate design and healthy balance of diversity has been unraveled across many regions of Africa, and unfortunately, local people are now paying the consequences for such loss. The implication for agriculture is that we need to go back to God. Since God delights in diversity, and because He designed creation and the Garden of Eden to be filled with diversity, our farming systems should be diverse. Multiple crops and multiple tree species for fruit, timber and firewood are the key elements that should be incorporated into as many of our farming communities as possible. God knew from the beginning that mankind would benefit from such diversity.

> Like God's garden, our farming systems should be diverse, incorporating many kinds of trees, plants, crops and animals

Principle 7

Godly agriculture strengthens the land and obediently strives to honor the sabbath principle

As highlighted in the introduction, farmers across East Africa are routinely expressing keen disappointment over declining crop yields. From the anecdotal evidence we have gathered, many farmers claim that current yields are only 25 percent or less of what they used to harvest. There are reasons for this. The big ones include the lack of soil conservation practices, resulting in erosion and the loss of soil fertility, and the simple fact that vast stretches of cropland have been repeatedly used, year after year, without reprieve. A famous agriculturalist and Christian, who also happened to be an African American, had this to say about such abuse:

> The farmer whose soil produces less every year is unkind to it in some way; that is, he is not doing by it what he should; he is robbing it of some substance it must have, and he becomes, therefore, a soil robber rather than a progressive farmer. We must enrich our soil every year instead of merely depleting it. It is

fundamental that nature will drive away those who sin against it. [1]

George Washington Carver

The typical pattern of agricultural practice that we see across much of Africa today is a pattern that continually takes from the land, without giving anything in return. Crops are harvested, the grain is used to feed people, and all crop residues are removed, either to feed livestock or to supplement the often meager supplies of firewood. The land is left bare, depleting organic matter within the soil, which in turn reduces its water-holding capacity. This pattern of unending extraction of all resources produced from the land leaves it exhausted and worn out. Once again, we have ignored the wisdom of God, and people are suffering as a result. Let us return to Leviticus 25.

> Speak to the Israelites and say to them: 'When you enter the land I am going to give you, the land itself must observe a sabbath to the LORD. For six years sow your fields, and for six years prune your vineyards and gather their crops. But in the seventh year the land is to have a sabbath of rest, a sabbath to the LORD. Do not sow your fields or prune your vineyards. Do not reap what grows of itself or harvest the grapes of your untended vines. The land is to have a year of rest.
> (Leviticus 25:2-5)

Another worthy quote from George Washington Carver highlights the timelessness and beauty of the Sabbath principle.

> We take this very Book, here - go way back here,
> almost to the beginning of time and we find, way back
> in the time of the Pharaohs, the farmers were obliged
> to rest their lands and every fifty years was a jubilee
> year. This was picnic time for the soil. [2]

God understood that the land needs rest, just like people need rest. The continued abundance of the Promised Land hinged on whether the Israelites were obedient to the sabbath principle. And in fact, God promised them a bumper crop in the sixth year if they obeyed this command. But the Israelites, in spite of God's warnings, did not obey, and they were guilty of the same thing that we are today. The consequences for the Israelites were disastrous. Among many acts of rebellion, part of the reason why God exiled the Israelites to Babylon was their refusal to give rest to the land.

> But they mocked God's messengers, despised his
> words, and scoffed at his prophets until the wrath of
> the Lord was aroused....He brought up against them
> the king of the Babylonians....He carried into exile to
> Babylon the remnant, who escaped from the sword,
> and they became servants to him....The land enjoyed
> its sabbath rests; all the time of its desolation it rested,
> until the seventy years were completed in fulfillment of
> the word of the Lord spoken by Jeremiah. (2 Chronicles
> 36:16-21)

Now this principle raises a serious and legitimate challenge for most small scale farmers today, many of whom cannot produce enough food for their families as it is, even if they cultivate all the land in their possession. When exposed to this concept it presents a virtual impossibility for many.

48

How can they afford to set aside a portion of their land to rest when hunger is staring them in the face? Are there other ways in which a farmer can strive to glorify God in rebuilding the strength and vigor of his land?

This is where I am deeply grateful for the ministry of Brian Oldreive, the founding director of a wonderful ministry called Farming God's Way. As a farmer of very distinctive success in Zimbabwe, Brian has developed a no-till method of farming that not only restores the strength of the land but has great potential to dramatically improve crop yields, even in drought prone areas. The key to the success of this technique is not simply to add petroleum based fertilizers, which is often mankind's quick-fix solution to our land degradation problems. Fertilizers do not address the roots of those problems, and in many cases farmers cannot afford to purchase fertilizers in the amounts that they require.

The key of Farming God's Way is a very deliberate attempt to rejuvenate the health and fertility of the soil, and it is based on something that we can observe in God's creation almost everywhere we go. With the exception of deserts, rarely does nature leave the soil of the earth to lie bare and unprotected. On the savannah it is covered by grass, and in the forest God covers the soil with a blanket of leaves. Farming God's Way attempts to replicate this example that we see in creation by covering the agricultural soil, at all times, with a layer of mulch called "God's blanket." This can include old crop residues, leaves, grass, and other plant material. It protects the soil from erosion and helps to conserve soil moisture, even during dry conditions. It also rebuilds soil structure and restores much needed organic matter, increasing the amount of water that the soil can absorb. In effect this reverses the process of degradation and exhaustion that afflicts so many landscapes today.

Brian makes an important point when he teaches on this technique: *"If we continue to extract food and grain from our fields year after year and never give anything back, our soils will become poorer and poorer until they can give nothing at all."* [3] The primary purpose of "God's Blanket" is to give something back, to feed and protect the soil, to revive and refresh its strength, so that in turn the soil will be able to feed us.

While there are many other biblical and technical ideals that embody the practice of Farming God's Way that deserve further explanation, it is this concept of giving back to the soil that enables farmers to fulfill the spirit of Leviticus 25:2-5. This is how we can honor God in our efforts to be obedient to the sabbath principle. If we do not have the ability to divide our land into portions, so that one of those portions can receive a periodic rest, then we must still strive to be as obedient as possible. For His sake, and for the sake of future generations, we must make every effort to care for the soil and renew its vitality.

> Godly agriculture strengthens the land and obediently strives to honor the sabbath principle

Principle 8

We bring greatest glory to God when our agricultural stewardship improves the fruitfulness of the land

N ow at first glance this principle may look rather similar to the previous one. While that is partially true, an important distinction needs to be made. Up until this point we have not outlined an explicit definition of agricultural stewardship, although the principles listed so far have clearly pointed in that direction. In a course I teach in a Kenyan seminary, I often ask the following question of the young pastors who are my students: "What teaching have you had on stewardship?"

In every case to date my students have answered in the affirmative saying "Yes, we have had teaching on the topic of stewardship." But upon closer inspection, you will find that "stewardship" has only been addressed in the context of tithing and the responsible use of funds within a church setting. Most of these young pastors come from rural agricultural communities, and unfortunately, the concept of "agricultural stewardship" is completely foreign to them. During a visit to Kenya a few years ago, Dr. Harry Spaling, Vice President of Academic Affairs at Kings University College in Alberta Canada, gave an insightful and concise definition of the stewardship concept, in the context of agriculture.

> Agricultural stewardship means caring for the land in a way which brings glory and praise to the Creator and which protects, preserves, and enhances the fruitfulness of the land.[4] (*my paraphrase*)

Let us pay special attention to the last statement in this definition. As agricultural stewards, standing before God, how can we best glorify Him? Is it by simply maintaining the status quo, by making sure that productivity does not decline, or should we strive to go beyond that by enhancing the beauty and fruitfulness of the land? While the answer here is rather obvious, we should take time to reflect on the Parable of the Talents to confirm our sentiments. Jesus is talking here about preparing for the second coming, and that as his followers, we should use whatever "talents" he has given us to the best of our ability.

> Again, it will be like a man going on a journey, who called his servants and entrusted his property to them. To one he gave five talents of money, to another two talents, and to another one talent, each according to his ability. Then he went on his journey. The man who had received five talents went at once and put his money to work and gained five more. So also, the one with the two talents gained two more. But the man who received the one talent went off, dug a hole in the ground and hid his master's money. After a long time the master of those servants returned and settled

accounts with them. The man who had received the five talents brought the other five. 'Master,' he said, 'you entrusted me with five talents. See, I have gained five more.' His master replied, 'Well done, good and faithful servant! You have been faithful with a few things; I will put you in charge of many things. Come and share your master's happiness. (Matthew 25:14-21)

We all remember the rest of this story. While the two servants who had used their talents wisely were praised and rewarded, the one who had done nothing was sternly rebuked, and what he had been given was taken away from him.

This parable gives rise to a very crucial question. How do Christian farmers view the land that has been entrusted to their care? Do we take the view of the man who was given just one talent, and do nothing to improve or multiply the productivity of our land? Do we simply view the land as a mere piece of property, bought or rented from someone else, to be used only as a means of putting food in our stomachs? Or do we look to the teachings of Christ and take a much higher view? Do we recognize that the land under our care belongs to our Master, that it stands before us as a gift that He has given and that His expectation is that we nurture and improve that land to the best of our ability?

As farmers we should be saying to ourselves, "This is my chance to demonstrate that I am prepared for the coming of Christ. When He arrives I want Him to look to me, and to my farm, and say, *"Well done, good and faithful servant."* Or alternatively, when the time comes to pass the land onto our sons and daughters, will they say "Thanks, father, for taking such good care of this land, and for how it stands as a testimony of your commitment to Christ. We are proud of you."

It is these kinds of sentiments that characterize the vision and goals of a Christ-exalting farmer. When we pursue agricultural stewardship in this manner, and succeed in our efforts to improve the fruitfulness of the land, we bring greatest glory to the First Farmer. Many years ago, the famed reforming pastor and theologian John Calvin pointed us in a similar direction.

> Let him who possesses a field, so partake of its yearly fruits, that he may not suffer the ground to be injured by negligence; but let him endeavor to hand it down to posterity as he received it, or even better cultivated. Let him so feed on its fruits, that he neither dissipates it by luxury, nor permits it to be marred or ruined by neglect....Let every one regard himself as the steward of God in all things which he possesses. [5]

We bring greatest glory to God when our agricultural stewardship improves the fruitfulness of the land

---

Principle 9

Like a good shepherd, farmers bring glory to God when they give proper care to domestic animals and livestock

In today's world our use and treatment of animals in the sphere of agriculture has been profoundly altered from the biblical norms that we see in Scripture. In many of our food production systems, animals are not seen as creatures created by God, but as machines to be finely tuned for maximum production, or as beasts to be employed solely for economic gain. Cattle are often crowded by the thousands into confinement feedlots, where they are fed unnatural diets of grain and high doses of antibiotics by automated systems, for the purpose of shortening the time span required to reach the slaughterhouse. Chickens and turkeys are raised in a similar fashion, often housed in small cages and long buildings, where they never spend a day outdoors in the reality of God's creation. There is something inherently disturbing about this picture.

While such industrialized approaches are less common in Africa, they exist and are becoming more prevalent, and apart from such "mechanized" approaches, we have our own set of problems in how animals are mistreated here. As just one example, a report compiled by Kenyans was issued in 2008 evaluating the quality of care given to donkeys in this country, which are commonly used for the transportation of various farm goods, water, etc. Interestingly, our home district of Limuru was given the lowest rating in the country. Animals and livestock in Africa are viewed in very much the same way that we view them in the developed world. Through our economically colored eyeglasses, they are seen as mere commodities, and they are treated accordingly. They are not viewed from a biblical perspective, as remarkable creatures created and owned by God.

Now some might start to worry that I am an advocate and supporter of the radical animal rights movement. That is not where I am heading. What I want to clearly point out, however, is that we as followers of Christ must legitimately assess this issue in light of a biblical worldview. If our forests, soils, fisheries, wildlife, and native plant communities have all been subject to mismanagement and heavy-handed abuse, and they have, then we can be quite certain that a similar pattern of abuse has been exercised in how we relate to domestic animals. Should a Christian exercise this type of cruelty or abuse? I think the answer is rather clear.

What is so profound in the Scriptures is the enduring concept of the good shepherd. Let us never forget that our Lord and Savior is powerfully and wonderfully portrayed as being the Good Shepherd. We are His sheep, and He does not treat us as a mere number among millions on the industrial conveyor belt. He is the Shepherd who oversees His sheep with unmistakable care. There is a striking beauty and richness that leaps out of this image, revealing an important truth. In biblical times a good shepherd was someone who cared for his animals, and he was someone who carefully adhered to and upheld certain standards. Let us review what surely stands as one of the most memorable Psalms of all time.

> The Lord is my shepherd, I shall not be in want. He makes me lie down in green pastures, he leads me beside quiet waters, he restores my soul. He guides me in paths of righteousness for his name's sake. Even though I walk through the valley of the shadow of death, I will fear no evil, for you are with me; your rod and your staff, they comfort me. You prepare a table before me in the presence of my enemies. You anoint my head with oil; my cup overflows. Surely goodness and love will follow me all the days of my life, and I will dwell in the house of the Lord forever. (Psalm 23)

David wrote this Psalm to illustrate how God is a Shepherd-King who pursues and supplies what is best for His sheep. A good shepherd has a keen recognition of the needs of his animals, and he provides them with food and water. He comforts, guides, and protects them from danger. And in Luke 15 we learn that a good shepherd even goes out of his way to rescue the one individual that has strayed and become lost from the rest of the flock. Christ died on the cross so that he might rescue His sheep.

There are many other examples from Scripture which serve to reinforce this concept. When God created Adam, he assigned him the task to name all the animals, pointing to the truth that He valued the animals because He was the one who created them. In the same way Noah was given the task of saving a remnant of all species from the flood, and God established a covenant of preservation with all creatures as seen in Genesis 9:12-13.

> And God said, 'This is the sign of the covenant I am making between me and you and every living creature with you, a covenant for all generations to come: I have set my rainbow in the clouds, and it will be the sign of the covenant between me and the earth.

As we move on in Scripture we read a very unique story in Numbers 22. Balaam, a pagan prophet, had set out on a journey with likely intentions to put a curse on the Israelites. The angel of the Lord stood in the pathway blocking him and the donkey he was riding. While Balaam was oblivious to the situation, his donkey sensed the angel's presence and stopped in its tracks. After beating the animal three times in the attempt to continue moving forward, Balaam was utterly surprised when the Lord enabled the donkey to give a verbal rebuke for such mistreatment. Later on in the book of Proverbs we read that the virtues of a righteous man can be seen in how he treats his animals. He is someone who is noticeably different and who adheres to certain standards.

> A righteous man cares for the needs of his animals, but the kindest acts of the wicked are cruel. (Proverbs 12:10)

Finally and importantly, when the Christ child arrived on earth, we see Him born in a humble stable, a place frequented by farm animals. The shepherds themselves are the first people to be informed about Christ's birth. And to this very day, in our celebration of the nativity, we commemorate this event with scenes of the Christ child surrounded by his parents, shepherds, and animals.

The conclusion here is unmistakable. God is concerned how we treat our animals. They take a special place in creation, providing us with uncountable benefits, and a caring respect should be afforded to them that reflects the goodness of God. How we treat animals is a testimony to our commitment and relationship with the Good Shepherd.

> Like a good shepherd, farmers bring glory to God when they give proper care to domestic animals and livestock

Principle 10

Godly agriculture works to bring justice and security to the poor and marginalized

As a child I lived in Ethiopia, and in recent years have had the privilege to return there and learn more about the challenges facing that country. My visits were prompted by an Ethiopian pastor who had attended one of our conferences. He stated very candidly, "If you think the environmental problems are bad in Kenya, you need to come to Ethiopia." I quickly discovered that his statement was quite correct.

A little over a century ago 30 percent of Ethiopia was covered with forest. Unfortunately today that figure has dwindled to less than 2.4 percent. During my most recent trip, as I departed from Addis Ababa on my return flight to Kenya, I carefully observed the landscapes and noted that we flew over mile after mile of farmland where there was hardly a tree in sight. Manure and crop residues, instead of replenishing the land, must now be used for fuel because firewood is so scarce in many parts of Ethiopia. And in terms of agriculture, recent estimates state that the country is producing less food today than it was 20 years ago.

The harsh reality of this was highlighted by David Shinn, former US ambassador to Ethiopia, who revealed that during the 24 years from 1980 to 2004, Ethiopia had suffered 15 years of poor rainfall. What was most striking, however, was the following statement by Shinn: Even in a year of good rains, Ethiopia still strugles with *"a structural food deficit of somewhere between four and six million people."* [6] Ethiopia is a land that cries out for God-centered creation stewardship and agricultural transformation.

In light of this critical situation, there is an often overlooked but highly important lesson we need to take from this country. Poor farmers in Ethiopia have often suffered from various forms of injustice and insecurity, and in any farming system a lack of justice and security inevitably cripples the progress that farmers might hope to make in the area of agricultural stewardship.

Let us imagine for the sake of example that a farmer decides to improve his land by establishing an agroforestry system. He works hard and spends extra time to seek advice and to learn some new things. He plants quality fruit trees near his home, and around the edges of his farm he plants several rows of various trees which he hopes will provide firewood or building posts to sell. After all this hard work the dry season comes, and one day, hungry livestock from a distant community are herded through his land. To his dismay they consume all the young seedlings which he has just established. He turns to his community leaders for help, hoping that they will be able to extract some form of compensation from the owners of the offending livestock, but no help is offered. There is no justice, and this leaves him disillusioned about making further efforts to improve his land.

Land tenure provides another good example. Countless small-scale farmers across the world today operate on land that they would love to own, but they often cannot obtain a title deed. They work season by season, with a great deal of uncertainty, not knowing if they will be able to remain on the land for years to come, or if they will suddenly be evicted. Without the security of ownership that comes with a title deed, even farmers who are committed to good stewardship will find it more difficult to spend the extra time and effort that is required to care properly for the land. The lack of justice and security often undermines and sometimes prevents farmers from doing the good things that are needed for maintaining productivity over the long term.

This was particularly true in Ethiopia during the socialist regime of the mid 70's to early 90's. As the story goes, many farmers actually had incentives to make their farms look worthless because of their fear of greedy government officials, who might come along and seize their property if it appeared to be healthy and productive. There are obviously many other forms of injustice that farmers face across our world. As Wesley Granberg-Michaelson once stated, *"Injustice has its roots in seizing and controlling part of creation for one's own selfish desires and thereby depriving others of creation's fruits, making them poor, dispossessed and oppressed.*" [7] The vision and practical expression of a godly agricultural system is that such a system will work to bring justice to the poor, the hungry and oppressed. God has a heart of compassion for the poor, as we are reminded in Isaiah 58:6-11.

> Is not this the kind of fasting I have chosen: to loose the chains of injustice and untie the cords of the yoke, to set the oppressed free and break every yoke? Is it not to share your food with the hungry and to provide the poor wanderer with shelter - when you see the naked, to clothe him, and not to turn away from your own flesh and blood? Then your light will break forth like the dawn......If you do away with the yoke of oppression, with the pointing finger and malicious talk, and if you spend yourselves in behalf of the hungry and satisfy the needs of the oppressed, then your light will rise in the darkness... You will be like a well-watered garden, like a spring whose waters never fail.

From an agricultural point of view this scripture captures something that is both deeply appropriate and beautiful at the same time. God begins by reminding us that we are to make a deliberate effort to show mercy and bring justice to the poor, the hungry, and the oppressed. When we pause and consider such people, what are the most common and urgent needs that they face? Often what they require most are the basic necessities of life, such as food, water, shelter, and clothing. Their first and primary needs, in many cases, are the good things provided from the "well-watered garden" of creation.

It is fitting to conclude, therefore, that Godly agriculture should be oriented towards protecting and ensuring that the "well-watered" garden of creation remains healthy and intact. This is one of the appropriate and important ways in which we can bring mercy and justice to the poor. And when we do this, God reminds us that in carrying out such efforts we become just like that well-watered garden. Our efforts are compared to a light that rises in the darkness, and we become "like a spring whose waters never fail." The farmers receive a blessing, and we receive a blessing. There is beauty to be found when we work to bring justice and security to such people.

> Godly agriculture works to bring justice and security to the poor and marginalized

Principle 11

Godly agriculture is not wasteful

Having spent half of my life in the United States, and half my life in East Africa, the topic of waste has always been a compelling one because of the clear distinctions that exist between two very different worlds. In terms of physical resources, the cup is overflowing in our western nation, and so are the landfills that receive the waste generated from our affluent lifestyles. By comparison to many other countries those of us from America cannot deny that we are a wasteful nation. Here in Kenya, we do a much better job, and when it comes to food, water, and energy, I truly admire the people for their frugality. Kenyans are generally not as wasteful with such resources.

But in all fairness, when we turn to a different resource, such as time, the picture is very different. A common phrase we hear in Kenya is often stated as follows: "We are running on African time," which customarily means "We will be late!" Time is a resource that is often wasted in Africa, leading to much frustration and delayed progress. When faced with such frustrations I truly admire my American culture for its commitment to timeliness, because time is a non-renewable resource. Once it is gone it can never be recovered.

Now when it comes to the agricultural context, how should we consider the general issue of wastefulness in light of our mandate to develop farming systems that glorify God? As a starting point we need to refresh our minds with the example that Jesus set for us in this area.

> Here is a boy with five small barley loaves and two small fish, but how far will they go among so many. Jesus said, 'Have the people sit down.' There was plenty of grass in that place, and the men sat down, about five thousand of them. Jesus then took the loaves, gave thanks, and distributed to those who were seated as much as they wanted. He did the same with the fish. When they had all had enough to eat, he said to his disciples, 'Gather the pieces that are left over. Let nothing be wasted.' So they gathered them and filled twelve baskets with the pieces of the five barley loaves left over by those who had eaten. (John 6:9-13)

What a wonderful precedent this is, modeled by our Savior, in a world where so many resources are wasted in some countries, and where so many lack even the most basic of resources in others. This example, if you recall, was repeated when Jesus fed the four thousand, where seven baskets of leftovers were collected. If we go back to Jewish culture we will discover that bread was perceived by the people to be a gift from God. In following with this perception, and if we rightly consider, we can agree wholeheartedly that we receive countless blessings from creation, all of which should be viewed as gifts from God.

To bring this concept closer to home a simple analogy may be helpful. Imagine if the president of your country, out of his own goodwill, stopped by your home to deliver a gift of food in a time of need. How would such a blessing be handled in your home?

We all know that if the food was used in a wasteful manner it would dishonor the one who gave it. And in this case the wastefulness would be a particularly flagrant violation because the gift came from your president. Now if we take this purely human example, and set our sights much higher, moving from a lowly human president up to the lofty and incomparable God of the universe, it raises a profound question: How much greater is the offense, how much greater is the disrespect that we afford to God when we waste the resources that He has given?

The conclusion is unavoidable. When we waste any resource, whether it be soil, food, water, energy, or time we offend the Giver.

A vision for godly agricultural stewardship recognizes that all resources on the farm must be handled with utmost care. Once again, this is where I appreciate the teaching of Brian Oldreive, previously referred to as the founder of the Farming God's Way ministry based in Zimbabwe. Brian points out that if we are going to transform agriculture we must be faithful stewards in all dimensions of farming. When everything is done on time, at a high standard, and without wasting any of our resources, we point to the excellence and perfection of God and we enhance His reputation among other people by our lives and action. This principle is linked very closely with the one that follows.

Godly agriculture is not wasteful

Principle 12

Godly agriculture pays attention to detail and pursues excellence in all aspects of its practice

If we can conclude that God is pleased by our efforts to not waste resources, we can also conclude that God is pleased when we pursue excellence in all aspects of our agricultural practice. While at first this might appear as being somewhat repetitive, I want to make a special distinction by focusing on the aspect of detail. Anyone who becomes friends with a passionate and successful farmer will soon learn that the mind of such a farmer is filled with details. For achieving the best results, he knows exactly:

- What type of crop varieties perform best
- How deep the seed must be planted
- The best spacing needed within and between rows
- How to apply fertilizer wisely and correctly
- The ideal conditions required for good germination
- When his crop is in critical need of a good rain to ensure that he will reap a good harvest

- Which weed and pest species he must contend with and when they are likely to cause problems

This list could go on and on. Such a farmer will know which part of his land is most fertile, how the water flows over it when it rains, and he will have a keen awareness of how well his crop is performing, along with a working knowledge of countless other factors. I have always been amazed at how an experienced farmer can be so detail oriented in the expression of his agricultural expertise. Such an expression points to the pursuit of excellence, and when we pursue excellence in the practice of agriculture, we bring glory to God.

So when Jesus states in Matthew 5:48, *"Be perfect, therefore, as your heavenly Father is perfect,"* we should not think it unreasonable to vigorously pursue the expression and application of high standards in our agricultural communities. After all, we serve a God who is very detail oriented, whose garden was beautiful beyond our knowledge, and whose creation speaks continually of His infinite perfections. He knows every bird in the forest (Psalm 50:9-11), when every sparrow falls to the ground (Matthew 10:29), as well as the very number of hairs on each of our heads (Matthew 10:30). And as Brian Oldreive reminds us, God even gave the Israelites very precise instructions on how to construct the temple and the Ark of the Covenant, just to highlight one of many examples in Scripture. Brian states *"I have found in my own life as a farmer that paying attention to details is very important. Success comes from being faithful, and from plugging away in doing all the little things well, and then you find beauty and effectiveness in the whole.*"[8] When we pursue agriculture with this type of commitment, we honor the First Farmer.

> Godly agriculture pays attention to detail and pursues excellence in all aspects of its practice

Principle 13

Repentance of our sinful and careless ways is the first and necessary step to bring healing to our land

What is most beautiful about the Gospel is that its very foundation is rooted in the beauty of forgiveness. We can admit our mistakes, we can confess and repent of our sins, and God is delighted when we approach Him in this way. He rejoices when His people are penitent, as we read in Luke 15:7, *"I tell you that in the same way there will be more rejoicing in heaven over one sinner who repents than over ninety-nine righteous persons who do not need to repent."*

It is instructive to remind ourselves how obedience and repentance in the Old Testament were linked to the abundance that the Israelites enjoyed from the land. Let us return once again to the Old Testament and read carefully what the Scriptures say in Leviticus:

> Follow my decrees and be careful to obey my laws, and you will life safely in the land. Then the land will yield its fruit, and you will eat your fill and live there in safety. (Leviticus 25:18-19)

> If you follow my decrees and are careful to obey my commands, I will send you rain in its season, and the ground will yield its crop and the trees of the field their fruit. Your threshing will continue until grape harvest and the grape harvest will continue until planting, and you will eat all the food you want and live in safety in your land. (Leviticus 26:3-5)

God had blessed His people with abundance and he had given them specific instructions to follow. His desire was that they would be both obedient and grateful as they partook of the bountiful blessings He had provided. Living in a right relationship with our God has innumerable benefits. In the book of Deuteronomy God expresses very similar sentiments. This time, as we see in the verses below, a stern warning is given to the Israelites. Showing disregard to God and the failure to walk in obedience to Him will have grave consequences.

> When you have eaten and are satisfied, praise the Lord your God for the good land he has given you. Be careful that you do not forget the Lord your God... Otherwise, when you eat and are satisfied, when you build fine houses and settle down, and when your herds and flocks grow large and your silver and gold increase...then your heart will become proud and you will forget the Lord your God, who brought you out of Egypt... (Deuteronomy 8:10-14)

> The land you are entering to take over is not like the land of Egypt... the land you are crossing the Jordan to take possession of is a land of mountains and valleys that drinks rain from heaven. It is a land the Lord

your God cares for; the eyes of the Lord your God are continually on it from the beginning of the year to its end. So if you faithfully obey the commands I am giving you today - to love the Lord your God and to serve him with all your heart and with all your soul - then I will send rain on your land in its season, both autumn and spring rains, so that you may gather in your grain, new wine and oil. I will provide grass in the fields for your cattle, and you will eat and be satisfied. Be careful, or you will be enticed to turn away and worship other gods and bow down to them. Then the Lord's anger will burn against you, and he will shut the heavens so that it will not rain and the ground will yield no produce, and you will soon perish from the good land the Lord is giving you. (Deuteronomy 11:10-17)

Much deserves to be said about these passages which we do not have time to explore. Apart from yet another striking portrait of agricultural plenty there are a few critical observations that must not be overlooked. Notice the caution in Deuteronomy 8. Satisfaction received from the fruits of the land is a good thing, but don't let that satisfaction become a distraction from our complete and continued dependence upon God. The bounty enjoyed from the land must always be joined together with an attitude that gives credit to the Provider Himself. Notice as well the explicit mention in Deuteronomy 11 that God cares for the land on a continual basis. If the Provider Himself cares for the land which brings blessing and satisfaction to our lives, then surely we too must care for that land. The Israelites were called to be good stewards, and so are we.

As we move on through this passage we also find the Great Commandment inserted into the very midst of the story, and this brings us to the final observation and main point. God wanted His people, an agricultural people, to remain centered on Him, in complete devotion and obedience. The bounty of the land would be withdrawn if the people ignored God. It is dangerous to forget the One who sustains and provides, and to assume the harvest of good things is the result of man's effort alone. If the Israelites turned away from God, and if they centered themselves on the counterfeit idols of the world it would only lead to their ruin.

Unfortunately, if we are honest with ourselves, this is exactly what our agricultural systems and farming communities have done today. We have forgotten God and we have left the First Farmer entirely out of the agricultural picture. Yes, it is true, when we go out into our fields we will often pray for rain, but that is usually where it ends. When it comes to our thoughts, our attitudes, and how we plan and carry out various farming activities, God is often far removed from the situation. This is why Christian farmers from all across Africa have little or no connection between their faith in Christ and their primary vocation in life. This is why many farmers willingly admit, "I don't know what my Christian faith means for farming." It is time that we put God back into the center of this picture, into His rightful place. When the Israelites put God back into His rightful place, when they came before Him with a spirit of repentance, God gave them a wonderful promise regarding their land.

> When I shut up the heavens so that there is no rain, or command locusts to devour the land or send a plague among my people, if my people, who are called by my name, will humble themselves and pray and seek my face and turn from their wicked ways, then I will hear from heaven and will forgive their sin and will heal their land. (2 Chronicles 7:13-14)

This verse offers great hope to those who struggle with hunger, and to those who are working to solve problems of hunger on behalf of the poor. As we watch the abundance that is fading from many of our landscapes, the Bible reminds us that our own sin lies at the very root of the problems that we face. We have been neglectful in our stewardship, we have forgotten God, and we need to repent of our sinful and careless ways. We need to seek forgiveness, and we need to plead with God. We must ask that He would open our hearts to be vessels of His grace, so that through His power we would be equipped and enabled to bring healing to the land. This is the first and most essential step that we must take if our desire is to bring glory to God in the practice of agriculture.

> Repentance of our sinful and careless ways is the first and necessary step to bring healing to our land

Principle 14

Godly agriculture gladly spreads new vision and success to others

The concepts we have been examining in this book hold tremendous significance and good news for the farmers of Africa. A fresh new vision for agriculture, rooted in the word of God, should not be hoarded within our churches, but spread far and wide, for the benefit of all people. In Deuteronomy 22:1-4 we catch a glimpse of God's expectations of the Israelites, who were called to exercise practical concern and care to the benefit of their neighbors.

> If you see your brother's ox or sheep straying, do not ignore it but be sure to take it back to him. If the brother does not live near you or if you do not know who he is, take it home with you and keep it until he comes looking for it. Then give it back to him. Do the same if you find your brother's donkey or his cloak or anything he loses. Do not ignore it. If you see your brother's donkey or his ox fallen on the road, do not ignore it. Help him get it to its feet.

Notice how this passage encourages us to view our fellow man as a "brother," whether they live next door or far away, and that we should even go out of our way to help such a brother. These words ring with the very same truth that Christ gives in the Great Commandment, where He instructs us to love our neighbors as ourselves. In following this kind of example our obedience to Christ helps to cultivate substantial benefits on the farming landscape. By assisting and serving their fellow man people learn the critical importance of trust, and how dependence upon one another can lead to mutual progress. With the building of trust comes the building of cooperative relationships and the community as a whole grows in its ability to solve its problems together.

If God is substantially present within our lives and within our agricultural communities, we will always be on the lookout to do that which is best for those around us. If God has blessed us with knowledge, with a new innovation or technique that improves the health and productivity of our land, then we will gladly share such success with the other farmers in our community.

> Godly agriculture gladly spreads
> new vision and success to others

Principle 15

Godly agriculture confronts impending challenges and plans for the future

A s I interact with farmers here in Kenya I am often humbled by the struggles they face, the discouragement they endure, and the reality that most have resigned themselves to passively accept the worsening trends that are taking place in many parts of the country. Agriculturally speaking, very few hold out hope for a brighter future, and even fewer demonstrate a sense of vision in restoring the land to its former productivity. As the body of Christ we have often ignored the spiritual and physical nature of the plight that such farmers face.

If we look to Ethiopia, we see that hunger and starvation has already become unacceptably common, and a nation of some 77 million people today is expected to grow to 160 million in the next 40 years. If we look to Kenya, a similar future is in store. By 2025 this nation of 37 million is expected to grow to some 60+ million. The need for healthy supplies of wood, water, and food is rising quickly, while the landscapes able to provide such resources are moving in the other direction.

How should the people of God respond to such problems that threaten the future of their communities? Have we taken this issue to the foot of the cross and asked "God, what is Your will for us as Your people who live on this part of Your creation? How can we encourage and equip Christian farmers to rise up to this challenge? These represent some very pertinent questions that we simply cannot avoid.

In many ways the future that currently stands before us resembles the very same future that once stood before a young man named Joseph. In Genesis chapters 37-41 there is an unforgettable story. As you will recall, God used Joseph, in spite of the many injustices that he suffered, to avert one of the most serious famines that we see in the Old Testament.

The Pharaoh of Egypt had a dream, where seven fat cows were consumed by seven gaunt cows, and where seven heads of lush grain were swallowed up by seven heads of scorched grain. This dream, interpreted by Joseph with God's help, foretold of seven years of plentiful harvest, to be followed by seven years of intense drought. The future presented a daunting challenge, but Joseph did not despair. This man of dreams, this man of character, integrity and self control, this man who had persevered under persecution, was also a man completely dedicated to God. And as a man who had humbly depended upon God his whole life, he was not one to shrink away from the challenges before him. In the case of the impending famine he became a man of vision who confronted the challenge head on. He proposed a wise plan of action to Pharaoh, and seeing his wisdom and integrity, the Pharaoh appointed him to put that plan into action.

As the story reaches its conclusion, we rejoice in seeing how God used a man like Joseph to save the lives of millions. We as farmers, and as Christian leaders in Africa, must become men and women who follow the example of Joseph. We must go before God and request that He would guide us, that He would give us a vision for the future, and that He would help us develop a decisive plan of action for our agricultural landscapes. And as we put that plan into action we must deliberately impart that same vision to our children and to others.

In considering the challenges that lie ahead, let us do everything possible to avoid making the same mistake that Adam made in the garden. When Eve was tempted by the serpent, he should have come running with a machete to protect and defend her from Satan's lies. Today we must reject the temptation to remain passive in the face of impending crisis, and accept our role as Christians, to honor God by depending upon Him as we take up our responsibility to fight for the future of our families and communities. This is the essence of what it means to be a Godly Christian in the agricultural context of Africa today. We must become men and women of vision, men and women of integrity, and people of initiative who lead courageously. In doing this we will bring glory to God and hope to the hungry.

> Godly agriculture confronts impending challenges and plans for the future

Principle 16

God-centered agricultural work is a mission field ripe for the harvest

As the son of missionaries Paul Brand had the privilege of growing up on the mountain slopes of South India when they were heavily forested. He later returned to India as a medical missionary to serve for eighteen years at the Christian Medical College in Vellore. In his excellent short story entitled *"A Handful of Mud: A Personal History of My Love for the Soil,"* Dr. Brand outlines his own experience with land degradation in India.

His early memories highlight the critical importance of soil conservation, a lesson taught by a prominent Indian elder, who scolded him and his local friends one day for damaging a carefully maintained terrace as they ran about catching frogs in a rice paddy. Paul never forgot this lesson, and when he returned to India many years later as a doctor, he was dismayed. Many of the forests were gone. The carefully maintained terraces where he used to play as a boy, and which used to provide rice, had completely eroded from their slopes, leaving behind bare rock in many places. As he travelled to other parts of India he noticed similar trends, and his heart became burdened for what was happening to the land and the soil. The following quote from his short story paints a clear picture of his sentiments.

> I would gladly give up medicine and surgery tomorrow if by so doing I could have some influence on policy with regard to mud and soil. The world will die from lack of soil and pure water long before it will die from lack of antibiotics or surgical skill and knowledge. [9]

Here we see a medical missionary, dedicated to God, who expresses the reality of his conviction in a gripping and concise way. Healthy communities are built upon the foundation of healthy landscapes. What is good for creation is good for people, and when the land is degraded, the people suffer from problems that medical technology cannot solve. Dr. Brand is not the only medical missionary who has developed a passionate concern for healthy land.

With over 35 years of community development experience in Central Africa, Dr. Dan Fountain, a Baptist physician, is another such missionary who has become a champion for agricultural and environmental stewardship. In his most recent book, *"Let's Restore Our Land, "* Dr. Fountain has outlined a simple but beautiful story of Pastor Simon, a rural church leader, who takes the initiative to lead his community in restoring the trees, forests, and soils of their land. Pastor Simon not only guides his community with principles from Scripture, but he is a true man of action. He is a farmer who knows the details necessary for improving the land, and he has implemented the very changes that he preaches. His example shows how Christ can work redemptively through His people to bring much needed transformation.

What these two examples illustrate is the need to restore something significant that has long been missing from the evangelical missionary paradigm. If our goal is to spread the gospel and to express the totality and fullness of Christ's love, then our mission work, whether initiated from foreign lands or from the local church, must move substantially in a direction that gives a meaningful response to the agricultural and environmental realities found in today's world. Not only is there a legitimate place for God-centered agricultural and environmental work in the context of missions, there is a desperate need for such work. For many evangelicals this represents a brand new frontier that needs to be explored, developed, and expressed. Bringing people hope for eternity and hope for today must be brought together into one cohesive whole.

In John chapter 4, verses 34-35, Jesus says:

> My food is to do the will of him who sent me and to finish his work....I tell you, open your eyes and look at the fields! They are ripe for the harvest.

What is God's will for evangelical missions in a world where chronic hunger persists in so many countries? Have we opened our eyes to the "fields" that lie before us?

The chance to transform both people and the land that sustains them is an opportunity that we must seize and a harvest that we cannot ignore. We live in a world today that literally cries out for godly men and women who can spread a passion for the supremacy of Christ in the context of agricultural and environmental mission work. In a hungry world nothing could be more beautiful, and more honoring to Christ, than to share the gospel with people, or to disciple them, while at the same time working with them to restore their land, so they might experience the joy of being able to feed their own families once again.

As my pastor, John Piper, once said, *"We should be living to meet the needs of others, because God is living to meet our needs."* [10] The fullest expression of the gospel must embrace both the spiritual and physical needs of people, and as this book has highlighted, there is a wealth of biblical truth that must be woven into the fabric of life surrounding the practice of agriculture. God-centered agricultural work is a mission field ripe for the harvest!

> God-centered agricultural work is a mission field ripe for the harvest

Note:
Dr. Paul W. Brand, together with Phillip Yancey, has co-authored two popular books by Zondervan: *"Fearfully and Wonderfully Made"* (1980) and *"In His Image"* (1984)

Dr. Dan Fountain's book *"Let's Restore Our Land"* can be obtained from ECHO (North Fort Myers Florida; 33917, USA, or from our Care of Creation Kenya office (Limuru, www.careofcreationkenya.org)

What if Jesus came to my farm?

As a final reflection let us imagine that one day, as you carried out your tasks on the farm, the Lord and Savior himself stopped by for a visit. What would Jesus do if he found you out in your field, working to remove weeds amongst your crop. Would Jesus sit down in your home, expecting you to stop your work, so that you might gather together the very best refreshments and serve him tea? Or is Jesus the kind of person who would gladly join you in the field, to assist in the task of weeding?

I have often presented these types of questions to my fellow farmers here in Kenya, which startles them at first, because they have never considered such a scenario. But these are good questions to ask. They help us bring Christ into our everyday lives. When considering the question if Jesus would join us in working on our fields, the initial reaction from many farmers is that he would never stoop to such a lowly task. But upon reminding them of a few events from Scripture, their eyes begin to brighten, and their perception of Christ takes on a brand new light. If Jesus was big enough to serve His disciples by washing their dusty feet, and if He expressed an interest in going to the home of an unpopular man named Zaccheus, and most remarkably, if He was willing to die on the cross for you, then yes, He is certainly gracious enough to join you as you carry out your tasks on the farm.

So these questions help to remind us of our winsome and loving Savior, who knows and cares about every detail of our lives. But if Jesus did in fact visit your farm one day, there would be other big implications to consider as well. As he joined you in the field, how would He evaluate the field itself, which is part of His creation, and which He has entrusted to your care? What would His conclusions be regarding this "talent" that came to you as a gift from His hand?

Would He be able to turn to you and say, "*Well done, good and faithful servant! You have been faithful with a few things; I will put you in charge of many things. Come and share your master's happiness,*" like we see in Matthew 25:21. Will this Jesus, the One who taught the parable of the sower, be able to say, "The heart of this man is like good soil. He has received the Gospel with joy, his life is transformed, and I can see evidence of that fact right here on his farm."

Like so many in today's world, let us not leave Jesus out of this picture of agriculture. As the Author of all creation, he was the one who originally put us in the midst of His garden. Let us now invite Him to come back into the midst of our farms, to take His rightful place, to stand alongside us, to guide and to teach us, and to mold us into the Christ-like farmers that He wants us to become. This is the glorious opportunity that stands before us, to transform agriculture, to glorify God, and to impart a two-handed gospel to the hungry; a gospel that that presents tangible hope for today as we also share and celebrate our eternal hope. Let us orient ourselves towards giving people a full and abundant life, following in the spirit of the words that Jesus gave to us in John 10:10: *"The thief comes only to steal and kill and destroy; I have come that they may have life, and have it abundantly."*

Conclusion

W hat we have seen in this book is a taste of the beautiful changes that need to take place within the farming landscapes of our world, and the benefits that such changes would provide to the poor. In order for such a transformation to occur we must examine this crucial dimension of life in light of the Scriptures and in light of the gospel itself. As Christians we should resist the temptation to be satisfied with the status quo. Neither should we be satisfied with simply providing relief supplies to millions of hungry people when drought or famine strikes. Our vision must be much larger than that.

Are we willing and committed to putting God into the very center of our thinking on this topic, so that everything we say, do and practice in the realm of agriculture brings glory to God and reflects our commitment to Christ? Are we ready to step forward and face the challenges of the future with a plan for action, like Joseph did hundreds of years ago in Egypt?

The task before us is clear. Where our agricultural landscapes have been degraded and rendered less productive, as Christians we must set an example to the world. We must work to restore and rebuild the strength and fruitfulness of those landscapes. Where our farming systems have damaged or continue to damage other parts of creation, we must lead the way in eliminating or minimizing that damage. We must demonstrate that we serve the God of all creation, and that we hold tightly to the standards set by a good shepherd in how we care for animals and how we care for the larger context of creation.

Where farming communities are discouraged and have lost hope in this primary vocation so common across Africa we must reignite that hope by encouraging those communities. We must bring justice and security to them where needed, and we must communicate clearly our agreement with the Bible, that farming is a respectable and noble way of life.

Together as the body of Christ we must collectively embrace a brand new vision for agriculture as a whole, one which demonstrates that we are following in the footsteps of the First Farmer. When we build agricultural systems modeled after the beauty and diversity that we see in the Garden of Eden, we honor God as the First Farmer, we brighten our testimony for Christ, and we move substantially in a direction that brings blessing to our brothers and sisters who are chronically hungry. Equipping and enabling them to feed their own families should be one of our top priorities. And in the process of doing this we can pass on a God-centered vision that will impact generations to come. This is how we can change the world, bring hope to the hungry, and glorify God in the practice of farming.

If Christianity is taking root in our hearts, then something should be taking root in our farming landscapes as well. If the gospel can heal and transform people, then surely the gospel can heal and transform how we farm.

Please join Care of Creation Kenya in casting this type of vision to Christians everywhere in Africa, and if you are involved in an evangelistic mission effort, please add this dimension of Christian living to those efforts.

Appendix 1
About Care of Creation Kenya

Care of Creation Kenya (CCK) is an evangelical mission organization dedicated to awakening the Church to its responsibility in environmental and agricultural stewardship. We are a registered NGO and our office is located at the well-known Brackenhurst International Conference Center near Limuru.

Our Mission statement
Our mission is to pursue a God-centered response to the environmental crisis in Africa which brings glory to the Creator, advances the cause of Christ, and leads to a transformation of the people and the land that sustains them.

Our Core Values:
1) The truth of God's Word and the power it holds to transform the lives of people.

2) The Lordship of Christ over all creation and over all aspects of life.

3) Obedience to and the fulfillment of the Great Commission and the Great Commandment.

4) God-centered environmental stewardship (the nurture, promotion, and practical implementation of a God-centered perspective and concern for all of His creation).

5) Holistic ministry (the healing and restoration of man's relationship to God, humanity, and the rest of creation; a ministry to the whole person, both spiritually and physically).

6) Exemplary and Christ-like concern for and treatment of people.

What We Do

Operating since 2003, CCK is working to address environmental degradation, and the hunger and poverty it causes, by discipling and mobilizing the Church to glorify God in caring for creation. Our efforts fall under these two main categories:

Promoting a God-centered Vision

o Spreading a biblical vision for creation stewardship by training churches, communities and institutions through local and national conferences and workshops.

o Developing and distributing biblically-based publications and brochures on creation stewardship.

o Partnering with churches and other organizations working to address the environmental issues of Africa.

Promoting God-centered Action

o Planting God's Trees
Advancing the development of a tree-planting culture in Kenya that brings glory to the Creator with a focus on indigenous forestry.

o Harvesting God's Water
Encouraging people to harvest and utilize God-given rainwater resources for life and land restoration efforts.

o Farming God's Way
Training and equipping farmers to protect and improve the productivity of their lands through a biblically-based approach to conservation agriculture.

Contact Information

Care of Creation Kenya (CCK)
c/o Moffat Bible College, P.O. Box 70
00220 Kijabe, Kenya

Office: +254.731.772203/04
Mobile: +254733451372
Email: craig@careofcreation.org
Web: www.careofcreationkenya.org

Care of Creation Kenya is affiliated with Care of Creation Inc., P.O. Box 44582, Madison, WI 53719 USA

Tel: +1.608.233.7048
Email: info@careofcreation.org
Web: careofcreation.net

Contact the US office for bulk order discounts or to make a donation to Craig's project in Kenya.

Appendix 2
A Declaration to Care for Creation

Care of Creation Kenya (CCK) has used the following declaration in its efforts to awaken and mobilize God's people towards a God-centered vision for creation stewardship. It was first signed in Kenya by more than 200 church leaders at the 2nd International Conference on God & Creation on March 11, 2006. Since that time hundreds of other leaders have also signed this declaration.

As leaders and members of the evangelical Church body in East Africa, representing a wide range of denominations and ministries, we stand together in agreement with the following declaration:

We believe in one God, the Creator, Owner, and Sustainer of all things, and we uphold the truth that His creation serves as a dynamic testimony of His power, wisdom, and glory.

As followers of Christ, we believe that God calls us to be good stewards of His creation. We embrace the truth that caring for creation brings glory to God, and that it serves as a practical expression of our love and concern for both current and future generations.

Upon reflection at this workshop, we believe the environmental crisis emerging in Africa poses a critical threat to our future. The creation is suffering as a result of deforestation, the degradation of agricultural and pastoral lands, pollution, the loss of biodiversity, and the greed of man. This is undermining the well-being of our communities and is leading to the impoverishment of our people.

We confess that the Church has responded poorly to this issue. Our failure in promoting and exercising proper stewardship over the creation has undermined our witness for Christ, and we hereby declare that we repent of our sin and negligence in this matter.

Acknowledging that the Author of our salvation is also the Author of all creation (Jn 1:1-3 and Col 1:16), we also declare that, more than any other group of people, it is believers committed to sharing the love and truth of Christ who should take the lead in responding to this crisis. We believe that awakening the Church to action is our most promising hope in the spiritual and physical battle against environmental degradation in the 21st century.

We therefore appeal to all Church and denominational leaders to recognize the gravity of the situation and to begin developing God-centered strategies to educate, disciple, and mobilize the entire Church to action. Our prayer is that God will initiate a powerful movement which will sweep across Africa and have an impact worldwide.

As we join together in a spirit of humility and repentance, and begin taking the necessary action, then we have reason for great hope! According to II Chronicles 7:14 this is the essential first step we must take if God is going to bring healing to our land.

"If my people, who are called by my name, will humble themselves and pray and seek my face and turn from their wicked ways, then I will hear from heaven and will forgive their sin and will heal their land."

Sources

1. Ferrell, J.S. *Fruits of Creation*. MN: Macalester Park Publishing. 1995, p 29.

2. Ibid., p. 30.

3. Oldrieve, B. *Farming God's Way Implementation Manual*, River of Life Westgate, Good Hope Farm. Harare, Zimbabwe. Local publication. 2006, p. 36.

4. Spaling, Harry. "Enabling Creation's Praise: Lessons in Agricultural Stewardship from Africa," *Biblical Holism and Agriculture: Cultivating Our Roots*. Ed. David J. Evans, Ronald J. Vos and Keith P. Wright. Pasadena, CA: William Carey Library. 2003, pp. 99-114.

5. Calvin, John. *Commentaries on the first book of Moses, called Genesis*. Ed. Rev. John King. Grand Rapids: W.B. Eerdmans Publishing Co., 1948, p. 125.

6. VOA News. (2006, March 8). *Ethiopia's Population Expected to Grow by More than 100 Percent by 2050*. Retrieved March 21, 2009, from VOANews@ VOANews.com: <http://www.voanews.com/english/archive/2006-03/Ethiopian-Population-Expected-To-Grow-by-More-than-100-Percent.cfm>

7. Granberg-Michaelson, Wesley. *A Worldy Spirituality: The Call to Take Care of the Earth*. San Francisco, CA: Harper & Row. 1984, p. 86.

8. Oldrieve, B. *Farming God's Way Implementation Manual,* River of Life Westgate, Good Hope Farm. Harare, Zimbabwe. Local publication. 2006, pp. 30-31.

9. Brand, Paul. "A Handful of Mud: A Personal History of My Love for the Soil," *Tending the Garden: Essays on the Gospel and the Earth.* Ed. Wesley Granberg-Michaelson. Grand Rapids, MI: William B. Eerdmans Publishing Co. 1987, p. 147.

10. Piper, J. (1993, August 8). *What Happens When You Die? Glorified and Free on the New Earth.* Retrieved March 24, 2009 from Desiring God website: <http://www.desiringgod.org/ResourceLibrary/sermons/bydate/1993/848_What_Happens_When_You_Die_Glorified_and_Free_on_the_New_Earth/>.

www.ingramcontent.com/pod-product-compliance
Lightning Source LLC
Chambersburg PA
CBHW050540280326
41933CB00011B/1654